Kids Nonfic
363.7 CURRAN
Curran, Lucy
E is for environment : the ABC
33410017466600 03-25-2022
S0-CPC-378

E IS FOR ENVIRONMENT

The ABCs of CONSERVATION

Written by

LUCY CURRAN

Illustrated by

FRANCESCA ROSA

RP|KIDS
PHILADELPHIA

For my A and my C. Always.—L.C.

For all our little heroes who help the Earth heal.—F.R.

Copyright © 2022 by Running Press Kids
Interior and cover illustrations copyright © 2022 by Francesca Rosa
Cover copyright © 2022 by Hachette Book Group, Inc.

Hachette Book Group supports the right to free expression and the value of copyright. The purpose of copyright is to encourage writers and artists to produce the creative works that enrich our culture.

The scanning, uploading, and distribution of this book without permission is a theft of the author's intellectual property. If you would like permission to use material from the book (other than for review purposes), please contact permissions@hbgusa.com. Thank you for your support of the author's rights.

Running Press Kids
Hachette Book Group
1290 Avenue of the Americas, New York, NY 10104
www.runningpress.com/rpkids
@RP_Kids

Printed in China

First Edition: March 2022

Published by Running Press Kids, an imprint of Perseus Books, LLC, a subsidiary of Hachette Book Group, Inc. The Running Press Kids name and logo is a trademark of the Hachette Book Group.

The Hachette Speakers Bureau provides a wide range of authors for speaking events. To find out more, go to www.hachettespeakersbureau.com or call (866) 376-6591.

The publisher is not responsible for websites (or their content) that are not owned by the publisher.

Print book cover and interior design by Frances J. Soo Ping Chow.

Library of Congress Cataloging-in-Publication Data
Names: Curran, Lucy, author. | Rosa, Francesca, illustrator.
Title: E is for environment: the ABCs of conservation / Lucy Curran; illustrated by Francesca Rosa. Description: Hardcover. | Philadelphia : Running Press Kids, [2022] | Identifiers: LCCN 2021001018 (print) | LCCN 2021001019 (ebook) | ISBN 9780762471706 (hardcover) | ISBN 9780762471683 (ebook) | ISBN 9780762499694 (ebook) Subjects: LCSH: Environmental protection—Juvenile literature. | Conservation of natural resources—Juvenile literature. Classification: LCC TD170.15 .C87 2022 (print) | LCC TD170.15 (ebook) | DDC 363.7—dc23 LC record available at https://lccn.loc.gov/2021001018
LC ebook record available at https://lccn.loc.gov/2021001019

ISBNs: 978-0-7624-7170-6 (hardcover), 978-0-7624-9969-4 (ebook), 978-0-7624-7168-3 (ebook), 978-0-7624-7896-5 (ebook)

APS

10 9 8 7 6 5 4 3 2 1

"I should be glad if
all the meadows on the earth
were left in a wild state..."

—**HENRY THOREAU, *WALDEN***

A is for ATMOSPHERE

The atmosphere is gases that surround the Earth.

B is for BIODIVERSITY

Biodiversity is the various life that makes up an ecosystem

C is for CONSERVATION

Conservation is the preservation and protection of the natural world—from animals and their habitats to water and the environment.

D is for DEFORESTATION

The destruction and clearing of trees and forests is known as deforestation.

E is for ENVIRONMENT

The environment is the natural world and all its living things.

F is for FLOWERS

Flowers aren't just pretty—they contribute to the air we breathe, provide food for birds and insects, and can even be used as medicines.

G is for GREENHOUSE

A greenhouse is a glass building in which plants are grown.

H is for HABITATS

The natural environment in which a plant, animal, or organism lives is known as its habitat.

I is for ICEBERG

An iceberg is a large floating piece of ice that has detached from a glacier.

J is for JUNGLE

A jungle is land covered by forests.

K is for KAFFIR LILY

Kaffir lillies are beautiful plants found in South Africa.

L is for LANDFILL

Trash and other waste are dumped into landfills as a form of disposal.

M is for MIGRATION

When animals move from one area to another with the change of season, they are migrating.

N is for **NATURE PRESERVE**

A protected area for plants and other wildlife is known as a nature preserve.

O is for OCEAN

Did you know oceans are the largest habitats on the planet and cover over seventy percent of Earth's surface?

P is for PLANET

A planet is a large celestial body that revolves around the Sun.
We live on the planet Earth.

Q is for QUARTZ

A mineral found in Earth's crust, quartz comes in many different colors, although it is clear in its purest form.

R is for RECYCLING

Recycling means reusing materials for other purposes and is one of the most important ways to help reduce waste.

S is for SOIL

Did you know that dirt and soil are not the same thing? Unlike dirt, soil is made up of living matter and minerals.

T is for TUNDRA

A tundra is a cold, treeless region or piece of land.

U is for UNIVERSE

The universe is larger than anyone can imagine and is made up of all the stars, matter, and living things.

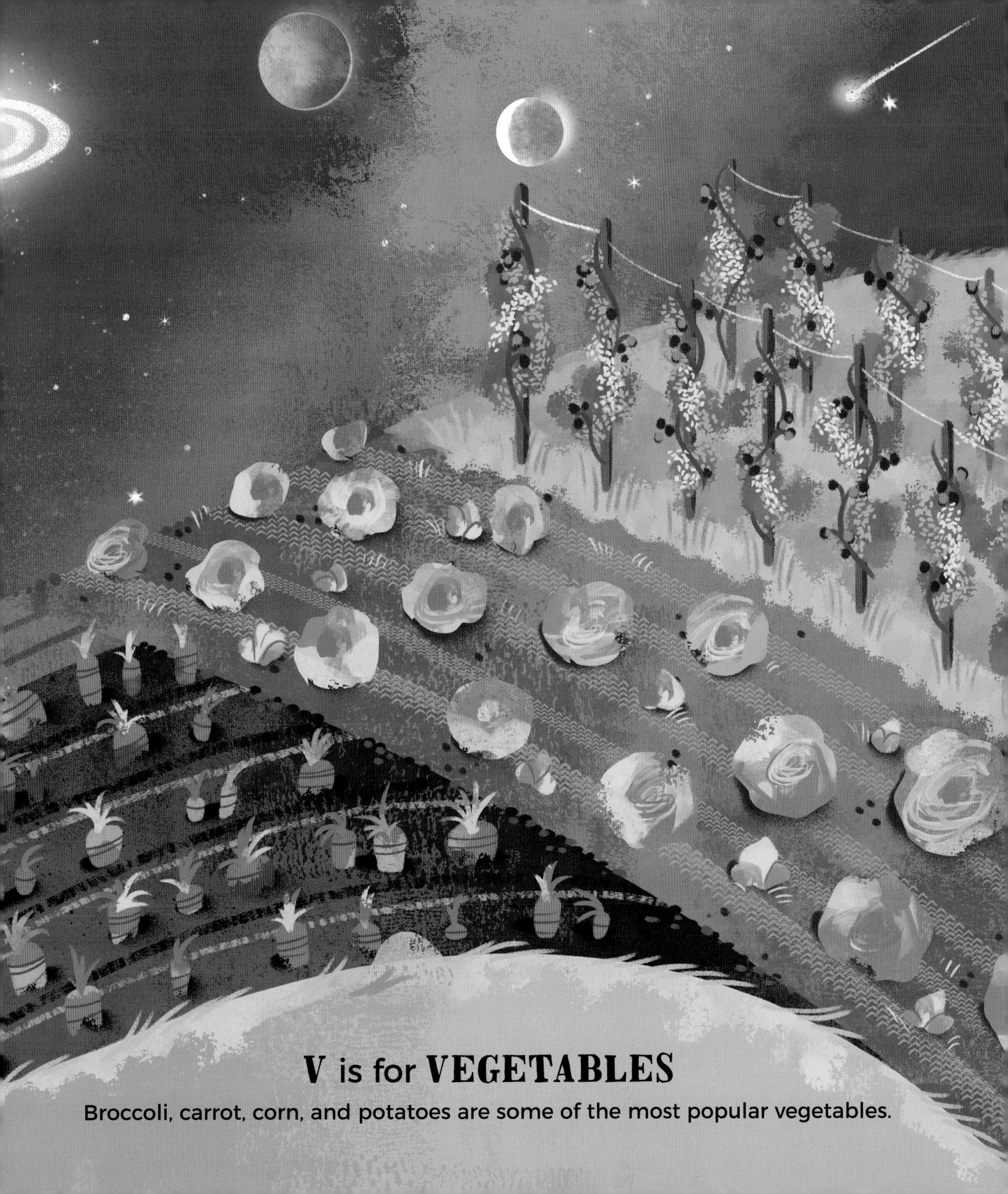

V is for VEGETABLES

Broccoli, carrot, corn, and potatoes are some of the most popular vegetables.

W is for WATER

Water has no color but reflects light to give it a blue or green or brown hue.

X is for EXTREME WEATHER

Linked to climate change, extreme weather is an event—snow, rain, drought, etc.—that is unusual for a specific place or season.

Y is for YELLOW-SPOTTED TREE FROG

Found in Australia, the yellow-spotted tree frog
is in danger of disappearing forever.

Z is for ZERO WASTE

The goal of zero waste is to reuse and recycle goods so that people are not creating trash that will be sent to a landfill.

We must all do our part to take care of our planet. From recycling and reducing waste, to planting trees and flowers, look for ways to help protect the environment and make this world a happier, healthier, more beautiful place for everyone!

The End